The mirror
of philosophy

The Mirror
of Philosophy

from the German alchemical
compendium

The Golden Fleece
or
the Golden Treasure
and Chamber of the Art

Translated from the 1604 edition

by Adam McLean

Alchemy Web Bookshop
2 Craighouse Square
Kilbirnie
KA25 7AF
U.K.

www.alchemywebsite.com

Now listen, you children of the ancient sages, and I will come and appear to you with the highest and most durable stone as is possible and reveal to you the state of human affairs and the most secret treasure of the whole world. I will also reveal it, not in a poetic way, nor in a mocking way, but in the most certain and friendly way. Therefore and for this reason let us listen with such devotion, as I will diligently convey to you the teachings of our Mastery, and then I will give you much better and more credible evidence of the things which I myself have seen with my eyes and touched with my hands, against all the entrepreneurs and landowners who have ever made use of much expense, toil and labour, and have not told you anything other than a hurried exit. Therefore I will tell you everything clearly and evidently, so that both the inexperienced and the experienced may understand something of this mastery, and so that no one may shame us for it.

Thus the ancient philosophers have written the matter in such an incomplete and utterly misleading manner that not only can their writings not be understood, but they also do not agree with each other, which is mainly due to the fact that they might deter those who aspire to such a noble art or its pursuit. But I, for my part, wish to remove all falsehood and obscurity, and to present to you the whole of the true experience, and therefore add the opinions of the wise, which best serve for such understanding, so that the whole matter of which we are speaking may be more clearly and fully understood.

Therefore, firstly we inform you that all those who work in this world apart from Nature are traffickers and vagabonds, and that they work in an unnatural matter, for nothing but a man is born of a man. So also from an unreasonable animal or creature only a beast is born, which thus gives birth to every thing of its own likeness.

Therefore, whoever does not have a thing from his own

1

nature may not have the same thing of his liking from another independent thing.

But this we say, so that no one may seek his fortune in vain; for if the operators have been deceived and brought to poverty because of their gross ignorance, they will also deceive others and bring them to poverty. I advise that no one should allow himself to be deceived in this art, unless he is prepared for incalculable expense beforehand; for the thing is most difficult to invent, and has made a fool of many for this reason. But once it is invented, you don't need much else, but only one thing, and not many more costs are involved, for it is only a few stones, one medicine, and one order, and this art is most certain, but the wise have known how to describe so many different colours, or the same order, for they have seen and heard it before.

Therefore we must make this statement, that all who work outside natural things are deceived, and that they are also tricksters. Therefore let your work be after the manner of Nature, so that it will be of value to you; then our stone is animal, growing, and productive; therefore be you of a good mind and will in the work of Nature, but do not now attempt to do it in another way. Then our art is not given up to several things but in one thing. It is now named in many different ways with many different names, however it may be, so it is always only a few things from a few different things. Nor is nature improved, but only in its own natural properties.

Iohannes Andreas speaks in the Book of Mirrors
about the title of the craftsmen.

You should know that the art of alchemy is a gift of the Holy Spirit and you should be aware that in our times we had Master Arnold de Villanova in the Roman Court, an excellent teacher and expert in the Holy Scriptures, of whom I have written above about the keeping of fasts, the *Cap: concilium*, etc.,

He was a great alchemist who made golden dishes and vessels, which he had proved in all tests: that no one should oppose or adopt this art, unless he had first prepared himself for it, and made sure that he would spend much money in it, for it is a difficult and secret thing to invent the matter.

The Mirror.

There have been some wise men who have said that this was the most terrible thing of all, with the consent of their will. I also ask the Creator of heaven and earth, who created all things, that what you are looking for is not bought for any money.

Arnoldus therefore says that it must necessarily follow that the active and the passive, that is, that the matter in which one acts and the active are both in common one and the same thing, but differ in form and image, just as a woman is different in character from a man, even though they are both human beings, yet they have in each other different properties and yet they are both human beings. It is the same with form and matter, that the matter must suffer that it is in it, but the form works in it and therefore wants to make the matter alike to it.

For this reason, matter by nature conceives the form or the formative image like a woman conceives a man, and as evil conceives good. Thus the body also grasps the spirit much more freely, so that it cannot come to any perfection, unless you know the natural root, then you will make the best much better out of it, then we may not call you by any way or means.

We will explain to you in any way or manner the name of this stone or call it by some other names, but by interpreting its parts we now give its appearance and its name, which is why our stone is called by the wise, a unique thing that contains within itself and from itself all things necessary for its perfection.

Arnoldus.

Therefore, it is understood from the same discourse that our stone is made of the four elements, and that the poor have it as well as the rich, and that it is found everywhere and is likened to all things. It is also composed of the body, soul and spirit; it also changes from one kind to another until the goal of its ultimate perfection. They have also said that our stone is made of one thing, and as this is true, then all our mastery is done with our water, and this water is the master of all metals, and all metals are dissolved in this water, as is evident. Then the imperfect body has been turned into the first water, and when these metallic waters are poured into our water and joined together they make a clear water which cleanses all things, and all the necessary parts are put into our water and all the necessary things are in it, and this is the most valuable and the worst of it. By this our Mastery is accomplished, then it dissolves the bodies, but not with the dissolution of the bodies, as the unwise say, who turn the bodies into the water of the clouds. It dissolves the body according to the dissolution of the wise, by which the body is turned into the very first water, and then the ablution takes place, and the same water turns the body into ashes.

Arnoldus.

The average experience shows the order between the different grades of the fire, that in the raising you should have a mild or soft fire but in the sublimation (that is in the raising or driving up) it should be moderate, in the purification or in the coagulation temperate, in the whitening firm, in the reddening strong.

If, however, in your ignorance you have erred in these matters, you will very often weep for the error and your labour.

Therefore it behoves you to be assiduous and diligent in your work, as understanding helps art, even as in the same way that knowledge helps art. Leave all other things aside and pay attention only to the completion of the work. Therefore that wise man Aristotle says, that the artists of alchemy should know that the forms of the metals cannot be changed unless they are perhaps transformed into their first original, in which case they may become unformed, i.e. they are changed into a different form and shape than they were before. This is why the regeneration or destruction of one thing is a transformation of another thing, just as much in works of art as in natural things,

4

for art works in Nature, and in some it improves and surpasses Nature, so that part of sick Nature is cured by the power of the physicians.

Therefore you should use the noble and venerable Nature, for Nature is not improved except in its own nature, to which you should not add or introduce anything foreign, neither powder nor any other thing, for various [and different] things or natures do not improve our stone, not even in that which is not of the original seed, for if anything foreign is added to this stone, it is immediately destroyed and is therefore taken out from it and has to searched for.

The Mirror.

That is why I want you to know, that unless at the beginning of the cooking of the liquid you add the same or similar things with some manual grinding, until all this becomes water, you will not yet have found the true work.

Then I add to you students' knowledge and make known to you this most delicious secret (since you do not trouble yourselves about anything else) that this mastery is nothing else, except that one boils the Mercury and Sulphur until they become one thing, then Mercury keeps the Sulphur from burning. Therefore the vessel must be well sealed, in order that that the Mercury may not disappear, nor the Sulphur may burn or be destroyed. Then the Mercury is indeed the very first water that everything that is sown into it does not burn because it has the water, no matter how strong the water may be, but if the water is consumed, it burns everything that is in the vessel

Therefore the wise men have commanded that the mouth of the vessel should be closed so that the blessed water does not leak out, but protects from burning what is in the vessel, but if the water is added to the same things, it will not burn the fire, and these things are created in such a way that the more the fire flames burn and work, the more the water is hidden from the innermost things, so that it is not injured by the heat of the fire.

The water carries these things away in its smoke and drives

away from them the flames of the fire. But I intend that all those who investigate this art at the beginning of the work should take a gentle, mild fire or smoke until the water and the fire may suffer each other, and as soon as you see that the water is at its best and will never rise or swell above you, then you must not worry any more about how the fire is created or how great it is. But it is good to keep the fire in good order until the body and the spirit become one thing, so that the bodily things become incorporeal and the incorporeal things bodily, after which we speak of the nature and kind of medium things.

Therefore the water is the thing that makes white and red. That water kills and makes alive, that water burn and makes white, that water cleaves and gathers together, that water decays and makes green thereafter and begets new things.

For this reason, my dear son, I charge you to use all your care to cook things properly, and do not let yourself spoil them, since you want the fruit otherwise, and do not let other useless things bother you, but pay attention only to the water, boil and simmer it finely and slowly, and ferment it until it changes perfectly from one colour to another. And take care that in the beginning of this work you do not burn its flowers, nor do you perform your work too soon, and be diligent that the door of your vessel is well closed, so that the one who is in it may not escape from it, and so, with God's grace, you will attain perfection.

Nature accomplishes her work slowly and leisurely, but I want you also to think and arrange your work according to Nature and see how and to what extent and from what causes the bodily metals are born out of the cooking of Nature in the veins and passages of the earth, and you should imagine this for yourself through righteous and natural perception as well as see out of what kind of things they come into being or rise above.

So, if you govern your work, you will find the perfect art, therefore keep this water in your hand along with its good effects, for it turns the white into white and the red into red.

The Mirror.

Therefore it is necessary that our stone be firstly extracted from Nature's two bodies and before a perfect elixir is made from the same stone, because it is necessary that this elixir be better purified and extracted from gold and silver, so that it should turn all perfect and imperfect metals into the gold and silver of the wise, imperfect metals into gold and silver, into the gold and silver of the wise, which they at least may not make perfect.

If the perfect metals give and communicate something of their perfection to something else, they themselves will remain imperfect, for the reason that they may not become whiter, for nothing is whiter than this, except so far as its whiteness extends. Therefore these things and works are done in our stone, so that the colour in it may be improved in its nature.

Also for this reason, an elixir be made of it, according to the hidden saying of the sages, from watery things some roots, a medicine, and a purification of all bodies shall be purified and presented as true silver and gold.

Here is the table of the great art of holy alchemy.

Firstly, in our green lion we have the warm matter of whatever colour it is and is called Adrop, Alot or Dunech Topem.

Secondly and thirdly, we have how the body dissolves itself in the Mercury, that is, in our Mercurial water, from which a new body is born.

Fourthly, we have the putrefaction of the Magi, which has never been seen in our times and which is called the Sulphur.

Fifthly, we have how the greater part of this water has become a black stinking earth of which all wise men speak.

Sixthly, we have how much of this black earth stood on top of the water at first and how much of it sank to the bottom of the water.

In the seventh, we have how the earth again dissolved itself and melted into water and changed into the colour of oil and was then called the oleum philosophorum, or an oil of the wise.

In the eighth we have how the dragon is born in the blackness and is fed with its Mercury and how it kills itself in it and drowns and the water begins to turn partly white and that is the Elixir.

Ninthly, we have how the water is cleansed from the blackness and remains a milky colour, just as many colours appear in the darkness.

Tenthly, we have seen how the black fog that was in the vessel on the water descended again into its corpse, from which it had come out.

Eleventhly, we have seen how the ashes have turned into the whitest colour, like a marble stone, and that is the elixir on white, and the child is the ashes.

Twelfthly, we have how the white has been turned into a translucent red like a ruby, and this is the elixir for red.

And if you want to understand the whole work well, read this book from one part to the other, and you will see wonders in our time. I have seen these things all the way to the ninth stage. I pray to God that he will grant me the grace to see this miraculous end. Following Arnold von Newendorff I believe that this work has been completed. I have seen all these things, and if I had not seen and grasped them, I would not have known how to describe them, nor, for that matter, how to write about them.

I have now revealed all the visible and necessary things in this work, but there are some things of which man cannot speak, nevertheless I have written it to a perfect conclusion and I know that such a work has never been written nor seen in a man's mind and soul. I bear witness to the teachers for this knowledge, that it is impossible to know such things unless from experience or from a teacher who teaches it, and you should know that this work takes the most difficult path, therefore it is necessary that one be patient and careful in this work and not be greedy or impatient.

The Mirror.

There are some fantasists who boast how they can make a gold potion or aurum potabile out of common gold or that they also believe that this gold potion is the best for healing and resisting all illnesses. There are also some doctors who boil ducats in water and say that this is the noblest and most useful for health, when it is in contradiction with their statement that this drinkable gold or water is not really the potable gold, nor is the water good or useful for health, but is evil and wicked to drink, then neither is common gold nor other metals good for health or healing.

But I may well say that such gold is good for the purchase of medicines and remedies and what also belongs to the sick to pay the doctor with. It is also good that he has a basin full of ducats to show the sick person.

Then it serves him very well that he should look at the gold, but the true gold potion or aurum potabile of the wise is the prepared elixir and the same is the aurum potabile, it is not visible, but more than capable of being used, for it is the greatest and highest remedy which expels all superfluidity, both from human bodies and from metals, for it converts all imperfect

9

metals into the very best and most precious gold, and in the same way it purifies all imperfect metals from all leprosy and disease. So also with the human body and this is therefore most certain and not at all doubtful and you can see that this is the opinion of all sages. But those fools who consider the common gold to be aurum potabile are blind and more then blind, because if the common gold gives something of its perfection to another thing, then that same gold will remain imperfect in itself.

It is made in two mountain trees

Her father is a virgin. His mother has not conceived. My dearest of all, we want to help each other and give each other a new son who is not like his parents, a king whose head is red, whose eyes are black, whose feet are white, and who is a master. Behold I come to thee, and am fully prepared to conceive a son, whose like is not in the wide world.

Thomas de Aquino.

But the matter of the stone is a coarse, thick water that flows there; the same water is hardened either by the water or by the cold; and you should believe that these stones are much more noble than those from the Turks, or those that come from elsewhere.

Lux luminum.

You will not be able to make any stone without the green and watery Duenach, which grows in our hearts.

The philosopher Rasis.

Son, visit the highest mountains on your right hand and on your left, and climb up to where our stone is found, and on the same mountain that bears and brings forth all kinds of picturesque colours and specimens is certainly our ore.

The philosopher Malchamech.

The stone that is needed in our work is of a workable substance; you will find it at all ends and places on the plains, on the mountains, and in the waters; the rich have it as well as the poor; it is also the most beautiful and the most beloved; it grows from flesh and blood; oh, how precious and noble it is to those who know it.

Here is the Solution of the philosophers and our quicksilver.

Let us go and seek the nature of the four elements which the alchemists bring forth from the fumes of the earth. Here we find the dissolution of the sages and from it we get our Mercury.

Our stone is unyielding, a light and durable substance that kills but also brings to life.

Mirror.

See that nothing foreign or repugnant is mixed with our stone, but set it alone.

Arnoldus.

Join our servant to his fragrant sister and they will perform the art with each other, then the white shining woman will be joined to the red man, so they will quickly heal each other and they will be joined together at the neck, and by themselves they will be detached and by themselves they will be killed so that the two will become one body.

Mirror.

You have asked how many are of the true colours. I will gladly tell you truly you should know that there are three perfect colours from which the others all take their beginning. The first colour is black, the second white, the third red. There are also many other colours on which one should not turn, then many other colours can be seen before the white colour.

The philosopher Hortulanus.

This describes the joining together of the two bodies of love, which is necessary for our mastery, and even though if only one of these two bodies were in our stone, it would still never produce some tincture.

Therefore it is necessary to join the two bodies together, and when the two bodies are joined together and have accepted or embraced each other in the joining together of the stone, then the stone is impregnated in the belly of the wind.

This is now what the wise man said, "the wind carried him in its belly". It is evident that the wind is the air and that the air is the life and the life is the soul, that is oil and water.

The philosopher Hermes.

I, who am exalted above all the circles of the world, have seen four faces, which have one father: the one face is in the mountains, the other in the air, the third in the stones, the fourth in the hills.

This stone is made of four composite elements.

Here the bodies dissolve entirely into our living Mercury, that is, into the water of our Mercury, and from this it becomes a permanent, best-quality water.

O Luna through my friendly gaze
You will see as beautiful as I.
O Sol you are above all love to recognise
But you need me as the cock does the hen.

Then you will see the nature of the four elements and find what you are looking for, but to invert the natures is so much that corpses are made into spirits in our mastery.

First of all, we make a coarse thing into a delicate thing, that is, from a body we make a water, and then from water we make a body, that is, the corporeal things become incorporeal and the incorporeal corporeal, and then we make what is below to be above and what is above to be below.

King Hermes.

When the bodies have been properly dissolved, they have been brought into the nature of the spirit and will never again be separated from one another, just as one water mixed into another can never be separated.

Hermes.

And indeed the whole work and government is nothing else than the water which contains in itself all the things of which we have need. Therefore keep in your hand this water with its good effects, for it makes that which is white, the white work, and that which is red, the red.

The philosopher Anaxagoras.

But there is one thing that contains the soul, the air, the calx and the four elements over which it rules, and it is not necessary to add to it other elements that do not conform to its nature.

The putrefaction of the sages, their Raven's Head.

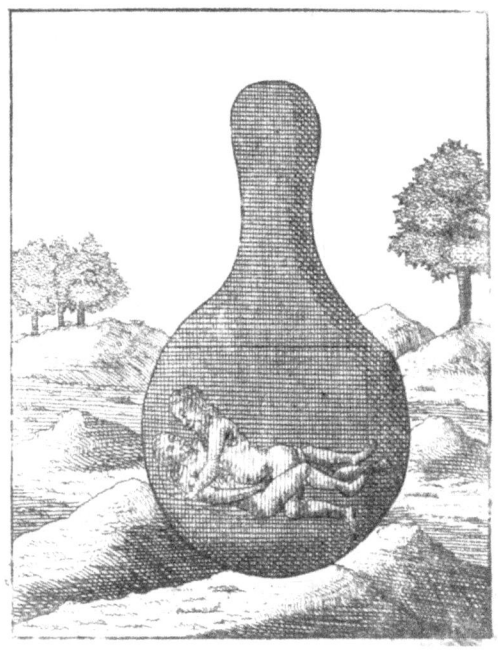

Putrefaction, nigredo, transponas [crossing over].

Penetrating and true black.

Here the bodies are in decay and become earth and when thou shall see the black then rejoice. This black is the beginning of your work.

It needs to become rotten.

The colour that appears after the black is flesh like.

Mirror.

Here the bodies are set in decay and become a black earth, and if you see the matter become black, then be glad, for this is the beginning of your work.

Arnoldus.

Heat our glass on a gentle fire until it becomes a body and the tincture is drawn out.

You should not, however, draw out such tincture together with another tincture, but it should come out a little and a little every day until a long time has passed.

The philosopher Hermes.

I am the blackness of the white, and the red of the white, and the yellow of the red, and truly, I speak the truth, and am no liar.

And you should know that the head of this art is the raven which in the blackness of the night and in the clarity of the day flutters without wings. The colour is taken out of the bitterness that is in his mouth, but the redness is taken from his body, and from his claws is taken the pure water.

The philosopher Hermes.

You should understand this and receive the gift of God and keep it from all the unwise, for it is hidden in the halls of the metals whose stone is irridescent, and it lies in shining colour in high mountains and a wide open sea. Behold, I have laid out all the figures for you.

Arnoldus.

But if it becomes black at the very first, we say that this blackness is a key to the work, for it does not happen without blackness. For black is the tincture or colour that we seek in this, with which we colour every body, which tincture is hidden in the metal, as the soul is hidden in the body of man.

Mirror.

Therefore, thou dearest son, when thou art in the work, be diligent that thou first have the black colour, and then thou shalt be sure that thou hast the right thing, and understandest the true way.

Geber, King and philosopher.

O blessed is Nature, and blessed is thy work, then thou mayest make of the imperfect a perfect one, with the true corruption, which is black and dark. After that thou makest new and many things green. With thy green work thou makest many colours to appear. In our work and mastery one should be patient and long-suffering, for evil comes from the devil in this art.

The Raven's Head

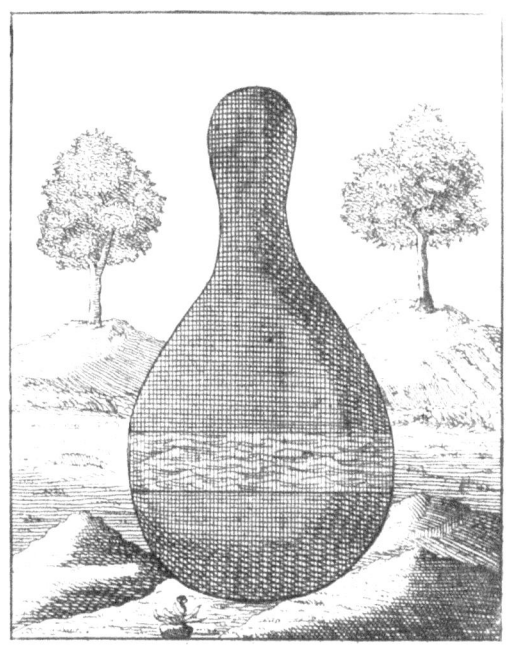

What is at the top of the picture is a dark mist. It is sea water or steam, the earth that is above the water will descend in the other vessel to the bottom of the vessel and grow worms in it.

The wise men saw still more that the matter became thick and turned into the earth, and that this thickening, in the beginning, occurred on the water, and thus as they let it slowly become thick, they saw that the earth sank into the water, and remained at the bottom under the water, which earth was yellow-coloured, black and stinking, they said, that this was the true corruption.

Mirror.

The fire should be diligently regulated in the oven, according to the instruction of the sages, so that it dissolves all matter into water. Then purify it with a gentle mild fire until the remainder turns into a black earth, which happens in twenty-one days.

Geber.

Dearest son, you should know that the art is nothing other than a perfect manifestation of God, then the whole of this mastery lies in a single thing, which we want to show you through the sayings of the wise, and according to the extent as we ourselves have seen and prepared it, with greater work and speed, and we have recognised that this single thing is perfect in itself, to the white and red.

Nor have we been able to find any other thing in which perfection, so far as concerns the perfect formation of the body, or the perfect preparation, would have to at first be completely and utterly destroyed and blackened.

Arnoldus.

Therefore be steadfast and persevering in the work, and in all things patiently persevere in the boiling, that the tincture go out upon the water into the black colour, and when thou shalt see the black go upon that water, then know that the whole body is melted and dissolved, then one shall in gentle measure, hold fast over this, until it receives the mist, which it hath given birth to, dark and gloomy.

The opinion of the wise is that the body, which has now become a dark powder, should enter into its water, and that it should all become one thing, then one water receives the other water as its own nature. Therefore it is that everything is turned into one water, otherwise you will by no means come to a perfect end.

The black Raven's Head.

The black earth from which snakes are born.

This is the black and stinking earth of the wise, in which worms grow, as one swallows the other, then one thing destroys another, then this earth is at the bottom of the vessel, and dissolves itself entirely in the water as before.

Here it is asked at what time the stone turns black, and what is the sign of its right dissolution?

I answer, if the black colour appears, then this black is a sure sign of the actual rotting and dissolution of the stone, but if the black disappears completely, then it is a sign that the stone has rotted and dissolved completely.

Arnoldus.

It is asked whether the same black fogs remain on the stone for forty days in succession.

The answer is yes, sometimes longer, sometimes shorter: this change and difference of the material and the size of the medicine, depending on how little or much of it there is, and also on how experienced the master of the work is, then a large part requires a long time, and a small part of the medicine requires a short time; the diligence and cleverness of the worker helps greatly to separate the black fog.

Arnoldus.

It is asked how long this rotting and cleansing of the earth lasts.

Answer, forty whole days, at times longer, at times shorter, depending on how much or little the earth is.

The Raven's Head is the oil of the wise.

Here is the fermented new black son, who is named Elixir. This black stinking earth, has turned into living Mercury, and has dissolved into an oily colour, therefore it is called the oil of the wise.

Arnoldus.

The gold, however, is dissolved so that it is brought into its first matter, that is, so that it becomes the true Sulphur and the living Mercury, and then we can make the best silver and gold.

If the gold is then changed into its silver and gold material, it should be washed and boiled so much that it becomes the true Sulphur and the living Mercury, then according to the opinion of the sages, the two things are the true material of all metals.

Hermes.

Whoever will take a wife and impregnate her, kill the body and bring it back to life, bring love and purify the face from blackness and darkness, will be the most diligent.

Then, when we add a gentle heat to our crowned king, our red son, and dampen him, they will become pregnant, and will give birth to a new red son, then their clouds, which were above them, will come back into the womb as they were before.

Arnoldus.

Hold fast to a persistent and temperate fire, until it dissolves into an impalpable water, and the whole tincture comes out in a black colour, which is a sign of the true dissolution.

The dark house.

Here the water begins to turn white to some extent, and the dragon eats his own wings.

And then it becomes a dragon that eats its own wings and emits various colours, then it will move in many ways from one colour to another until it arrives at a permanent colour.

The philosopher Aristenes.

The most joyful animal should not be fed, let it be hungry and thirsty, and know that it will remain hungry and thirsty for three days.

Here is born the dragon, and his house is darkness and blackness, in all these he dwells, over that same sea death and darkness fly away, and the dragon makes the sun's rays perfect. Our dead son comes to life, and the king will come out of the fire, and will be happy to be married together, and the hidden things will be revealed within. The virgin's milk will be made white, and our son will now be brought to life, and he will be an unsurpassable master of the tinctures.

The dark house, the ascending of the wise men.

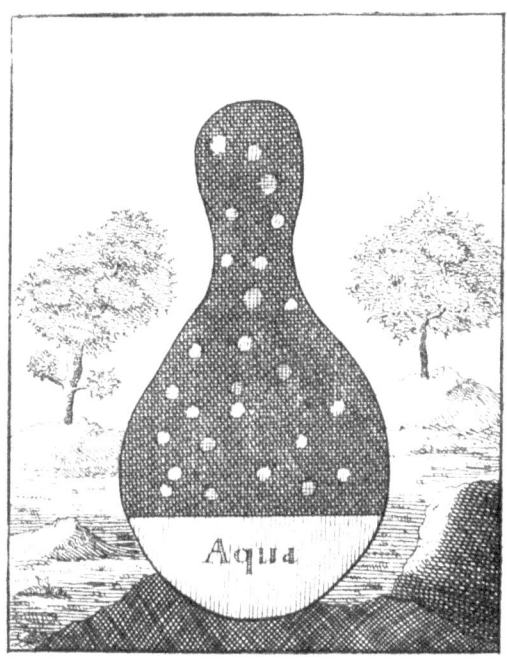

Here he purifies himself from the black, and becomes white as milk.

Take the black, which is blacker than black, then many and various colours will appear in it, and the virgin's milk will be whitened.

Hermes.

And when our Sulphur is alive, he will be a man battling against fire, and will excel in tinctures.

Arnoldus.

Out of the sea the clouds rise, and then rain descends upon the earth, then every heavy and thick body falls down upon the earth.

But the living Mercury supplanted from the earth, whereof all things are made, is a pure water and the true tincture, which hath divided the matter, and that is the white Sulphur, which alone whiteneth the metal, whereby the spirit is preserved that it cannot fly away.

The teacher of this book.

A spiritual person has had these words in the events of the revelation of our Lord Jesus Christ, which have never been heard by anyone, or found written in books, so you should know that the neck of the vessel is the Raven's head, which you must kill, out of it a dove is born, and after it the Phoenix, ever since happy.

All this is understood of the entire Mastery, namely White and Red, in these few words.

The ash of ashes.

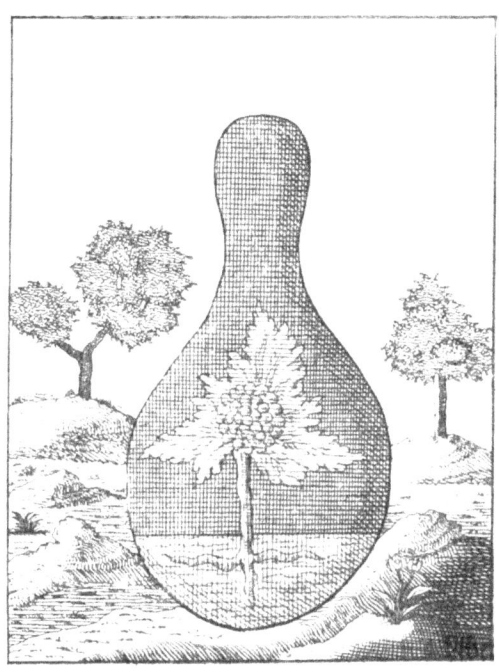

The black mists have descended onto their bodies, from which they have risen, and have become a union between the earth and the water, and have become ashes.

But because Nature does not move, then only by the action of the works, therefore if you measure the works well, then fire and water will obey you, then they wash away the body, purify, and increase, and take away its darkness, the water dwells in the air, follows the earth, as iron follows the magnet.

Arnoldus.

Therefore you should work all this order of preparation over him for the fourth time, and the last time calcine him in the most perfect and calcined way and manner, then you will have the most precious and noblest earth of the stone, very well regulated by a noble hand.

But to calcine or turn it into calx is nothing else than to dry up the matter and turn it into ashes.

Therefore burn it without fear until it becomes ashes.

If it has now become ashes, then you have mixed it well; you should not regard these ashes lightly, but you should return to it the sweat that it has exuded.

Arnoldus.

When the water has been completely excreted, and has been absorbed into the earth, it should be left to cool in its vessel for several days, until the noble white colour appears.

In this vessel all the colours of the world will appear, and the moisture will be extinguished.

Arnoldus.

The water that has come out of it, give it back to it until it becomes perfect and cannot separate from it in the fire, that is, let the black that has separated from the body be brought back to the body from which it came out and become a corpus.

The White Roses.

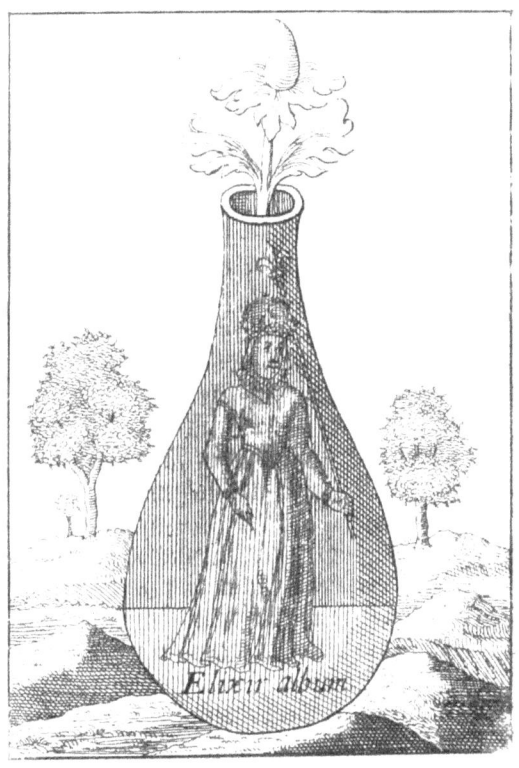

I am the elixir of whiteness, which transmutes all imperfect metals into silver, which is better than the silver from ore.

One part of this elixir converts a thousand parts of living Mercury into the purest silver.

Arnoldus.

Make the laton white, that is the body, and put it again, break the books, lest your heart be broken, then our matter is small, also requires little help. He who made me white, make me also red, white and red both spring from one root, what happens in the white, happens also in the red.

Therefore, my son, if you work white, and do it diligently, and do not overshoot the goal in your work, blessed are you, if you understand this properly, then wonder, fear, and fright will impel you, boil, grind, and cook it again, do not let yourself be distressed, and try again, whether the whole work is long, then it will be accomplished by long cooking.

Hermes.

You should know that the sun's flower is the stone of stones, therefore roast it day until it becomes like a glittering marble stone, and know when it thus shines, then you have the great secret, then there is a stone that surpasses all other stones.

Arnoldus.

You have learned to make the white, my dearest son. But now there is a promise of the red, but still, if you have made the white first, then the true redness cannot be made, then no one can come from the first to the third, then only through the second.

Thus you will also not be able to get from the black to the gold colour, but only through the white, because the gold colour is made of many white colours, and the purest black colour is added.

Therefore you should make the black colour white and then you should colour the white colour red, so you have the mastery. Just as our year is divided into four parts, the first part is the cold, damp and rainy winter, the second part is the warm, damp and blossoming spring, the third is the warm, dry and reddish summer, the fourth is the autumn, cold and dry, a time to harvest the fruits.

With this sequence you shall govern the tinging natures, until they bring the rich moisture with them. But now the winter is passed, and the rain is come and gone, and the flowers have appeared in our land in the time of spring.

But we have rested on the white roses, as they have been, which have changed the black body into a bright silver.

Therefore, when you see this white appearing, exceeding all of them, you can be sure and certain that the redness is hidden in this white, then you should not take this white out of the oven, but should boil it until the whole material turns red.

The Red Roses.

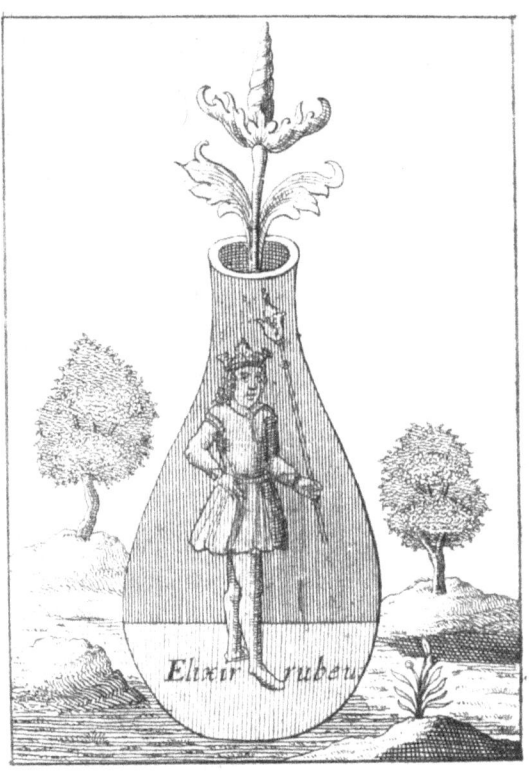

I am the elixir for the redness, I change all imperfect bodies into the purest gold, which is better than the gold made from the ore, and if one throws a part of this elixir on a thousand parts of the living Mercury, we have found that it has coagulated it and made it red, that it has been changed into the very best, purest and most valuable gold.

The philosopher Nillius.

In the end the Lord will go forth, the King crowned with his royal crown, shining as the sun, clear as the sparkle of a diamond, as fluid as wax, as persistent in fire, penetrating and possessing the living Mercury.

Arnoldus.

The red colour comes from perfect fermentation, for blood does not grow in a man unless it is first well cooked in the liver, as also we, if we get up in the morning and see that our urine is white, know that we have slept too little, but if we go back to bed, and then sleep again, our fermentation is completed, and our urine turns yellow. In this way also, the white may come to the redness by cooking, so that the fire is ripened, and since our white mass is also cooked diligently, it will become the very best red.

Hence come these words: The white shall not be red, but shall give its light appearance, shall not boil, but shall rest in a constant heat until it has its appearance, until it takes on greenness, and out of white comes the red colour.

Therefore it should be boiled in a dry fire and dry calcination until it becomes red like cinnabar, after which you should not add water or anything else to it until it has boiled down to its perfect redness.

Nillius.

If it is true that it has become red through further cooking, it will give a colour of everlasting gold. There is a herb which is called Adrop, Duenach, Alot, from which this medicine is made, and is found on a high mountain in the field of N then, in the number of the seven days all work is comprehended, thus in the seventh number everything is formed and made perfect.

This work is often done again, and is often distilled through seven week days, so that the days remain in order.

This is the property of our stone.

He also has a subtle virtue to heal all kinds of sicknesses above all other kinds of sicknesses, then he pleases the mind, he heals virtue and puts an end to old age, he does not let the blood rot or take over, nor does he make the colder parts burn, or tear down the melancholia, but he increases the blood beyond measure and cleanses it.

He cleanses what is in the spiritual members, and all bodily members he restores to health, and protects them from harm, so he heals all heat and cold diseases in the most effective way. If a disease has lasted a whole year, it will heal in twelve days, but if it has been a disease that has lasted a long year, it will heal in a month and be cured in a short time.

It casts out all evil moisture, and makes good moisture in its place, it gives favour and honour, whoever carries this stone with him, it gives him security, strength and overcoming conflict, and in this the highest secret is accomplished, over the beauty of the whole world, a most noble treasure, to which nothing can be compared; to him who has it, God reserves his mind, that he may not reveal it to the ignorant, or that it may be made manifest.

This book is called the Gift of God.

And thus is accomplished the noble gift of God, which is a secret treasure, and an incomparable possession, then, as Plato says, he who has this gift of God has the kingdom, or the wealth of the whole world, then he has come to the end of riches, and has broken the bond of Nature, and not only because he is able to transform all imperfect bodies into the purest gold and silver, but most of all because he is able to protect and preserve man and every other animal from all diseases.

The crystal metal, which is the white elixir, when given in the form of a large mustard seed to a feverish person, cures him, and if a leper is cleansed with the same metal through the four seasons of the year, he is cured and cleansed of the leprosy. And if someone has the evil leprosy called elephantiasis, he is cured with the red powder of which saffron is made twice a year, namely in the months of March and autumn, he is cured, both the red and the white powder makes him healthy. It also cures the paralysis, the falling down in danger of death. And if the powder is given to a woman in childbirth, it will ease her delivery.

But Geber says that the red elixir heals all long-lasting illnesses that have despaired the physicians, and brings a man back to life, so that he becomes as beautiful as an eagle, and lives for five hundred years, and even longer than some wise men have done, because they used such elixirs three times a week, in the quantity of a mustard seed.

There is a herb called Saturnus, from the root of which is made such a medicine, then every like sticks to its like, every form frees itself from its form, every sex frees itself from its sex.

Therefore notice that all diseases that arise from the top of the head to the soles of the feet, and if the disease is as old as a month, it is healed in a day, but if it is a year old, it is healed in a month, then because it heals all imperfect metals from all diseases, it also heals all human bodies.

Therefore our appointed stone, not unreasonably is called the great Theriac, heals the bodies of men as well as metals, of which Hermes, a king of the Greeks, and a father of the wise, says.

"If you take of our elixir every day for seven days as much as three grains of mustard seed, your grey hair will fall from your head, and your black hair will grow in its place, and so you will turn an old man into a young one."

Hermes.

This book is called the Book of the Composition of Alchemy, what Alchemy is, and what its composition is. The Latin language has not yet recognised this name and composition of alchemy; in the present discourse it uses this unknown obscure name, under the interpretation of clarity. Hermes, however, and other wise men after him, expound this name and say in the Book of Transformation that alchemy is a bodily essence, composed of one and the same thing, which combines the noblest things with each other through art and workmanship, and transforms them with natural blending into a better substance.

Thomas of Aquinas.

The transformation of metals is such that the essence of one thing is changed into the essence of another thing, although Aristotle and Avicenna say that the artists of alchemy should know that the form of metals can never be changed, but it is written afterwards that the same form is then brought back into its first matter, this matter being the closest and most similar to all metals, especially to the living Mercury. However, in the first place, this matter is water, but such water is very much like and similar to nature, even as natural, although Nature is helped by art. And through these difficult things many have become fools, who through this art have lost their youth and their property,

There are so many kinds of work, which have been described by these untrustworthy people, that if you spent all your property on this work, you would despair of the art and never obtain the desired thing.

But considering all things, as kings have had subtle workers, who afterwards hardly or never attain the perfect end.

I have thought that art was not true or just, but I have gone into it myself, have looked at the books of the mysteries of Aristotle and Avicenna, and have found them to be true, although they have therefore confused and wrongly explained them.

I have also examined the books of their opponents, and have also found such contradictions. I have also examined the natural causes, and found that by them the art might be performed. Then I have seen that the living Mercury passed through the metals, and that other metals grew out of it.

Therefore I suppose that no one would refrain from approaching the work for the sake of the things that are yet to be told, he is then well experienced in the natural beginnings, and well practised and skilled in the ways of distillation, coagulation, dissolving, and most of all in the government of fire, that he is also not a man who wants to hurry, but who wants to be slow, and who wants to act with care and prudence.

It is a stone of the body, of the soul, of growth, and of the ore, out of which is drawn the virtue of the white colour, or of the pure redness, that they add nothing, and is drawn out by the separation of the four elements, their purification, and their assembling together in the name of the Lord.

Take a pound of it, grind it small, set it on a marble stone, and mix it with a pound and a half of common oil, and it will become like a dough. If you throw it a little on some of the metals, it will change them into silver or gold very easily. There the four elements are hardened and purified, some with a spirit and body, and though I have not baked this work, yet I perceive it to be natural and true, believing also that it is well known to all.

But Avicenna in *Epistola ad Hermetem*, or N. his cousin, describes this work of several rnasses, and if I had had time, I would also have described this work properly.

Mirror.

But that you may understand the true interpretation the more clearly, we wish to indicate to you at what hour and on what day and also in what month you are to begin our mastery, and we say that whoever works otherwise will certainly suffer and be lacking, and whoever works according to our commandment will certainly see the true righteous art and the true mastery.

Therefore we command that one should take the Philosophers' Stone with all its essence and take out of this stone its purest and most subtle essence. One should put it into the vessel of the sages and seal the mouth of the vessel after the manner of the sages and put it into the furnace of the sages when the sun goes down.

But all this must be done on a Monday, in the middle of the month of Christ (called December) and be continued until the half month of January under the sign of Capricorn, and then one must set fire to the Magi's fires through the whole sign and govern the work according to the use of the Magi, and be

41

diligent that in the sign of Capricorn all the volatile matter may be thoroughly perfected.

Mirror.

The heat should be such that you can hold your hands unharmed between the sides of the stove, in this heat let it stand until you see the matter begin to turn black, and if it is too long a delay that it does not want to turn black, then stiffen the fire a little, then you will see that the matter will turn black, and be glad. This then is the beginning of the change, then keep the fire in a steady state, until all colours will be gone, and when you will now see the matter become white to a certain extent, then increase the fire gradually in a steady way, until it comes to a perfect white colour, then it has enough of it, and the white is completed to the white.

But the fire should be regulated very slowly, according to the indication of the matter, until it rises to the white colour.

Geber in his book, called Summa, in the 16th chapter, begins with natural things.

It is to be noted that after our stone has been purified and completely cleansed of all destructive things, and after it has been cleansed, you do not need any further purification of the vessel, nor do you need to open it, but only trust that God will keep it from breaking.

And therefore the wise men have said that in one vessel the whole mastery shall be accomplished, and it is to be known that in forty days, and in so many nights, it shall be completed.

The work on the white colour, according to the true purification, cannot be determined in time, but only as the artist performs it well, and in sixty or ninety days is accomplished to the redness, or even so in ninety-one days, until the very end, and this are the true goal and time for the entire completion of the work, and if you have come to this, then praise the Lord Jesus Christ. Amen.

Alchemical Translations Series

107. The Golden Mirror of Outer and Inner Vision
108. A Hermetical Banquet
109. Addresses to the Gold- and Rosy Crucians - Ecker
110. Chrysopoeia - Augurel
111. The True and Perfect Preparation - Samuel Richter
112. The Hellish Goddess Proserpina - Rudolph Glauber
113. The Powder of Projection - D.L.B. Lord of la Borde
114. Gold Unclothed - Johann Christian Orschall
115. The Divine Arcana
116. Three Curious Alchemical Writings
117. Theory and Practice of the Gold and Silver Trees
118. The Philosophical Water
119. A Treatise on Metals and Alchemy - Bernard Palissy
120. Chemical Essays - Karl von Eckartshausen
121. The Aurora - Henri de Lintaut
122. Of Hyle, the Universal Prima Materia - Khunrath
123. The Four Amphitheatre Engravings - Khunrath
124. The Fire of the Magi - Khunrath
125. The princely and monarchical Roses of Jericho - Fictuld
126. The Oraculum Manuscript
127. Kings of Scheschian
128. The Theoricus or Second Degree of the Rosicrucians
129. Nine treatises on Goldmaking - Stephanos
130. The Philosophers' Stone - Athanasius Kircher
131. Aelia Laelia Crispis - Nicolas Barnaud
132. Sphynx Rosacea - Christophorus Nigrinus
133. The Theory of the Divine Art of Alchemy - Mylius
134. Summum Bonum - Robert Fludd
135. The Philorcium of George Ripley
136. The Hieroglyphics of the Egyptians - Michael Maier
137. Hermes - Emblems of the Twelve Nations - Michael Maier
138. Maria the Jewess - Emblems of the Twelve Nations - Maier
139. Light emerging by itself from the darkness
140. The portable laboratory - Johann Joachim Becher
141. The Hieroglyphics of the Greeks - Michael Maier
142. On the Creatures of the Aethereal Heaven - Robert Fludd